HOW CAN I EXPERIMENT WITH ... ?

FORCE & MOTION

D1314170

Cindy Devine Dalton

Cindy Devine Dalton graduated from Ball State University, Indiana,
with a Bachelor of Science degree in Health Science.
For several years she taught medical science in grades 9-12.

Teresa and Ed Sikora

Teresa Sikora teaches 4th grade math and science. She graduated with a
Bachelor of Science in Elementary Education and recently attained National Certification
for Middle Childhood Generalist. She is married with two children.
Ed Sikora is an Aerospace Engineer, working on the Space Shuttle Main Engines.
He earned a Bachelors of Science degree in Aerospace Engineering from the
University of Florida and a Masters Degree in Computer Science from the
Florida Institute of Technology.

Rourke Publishing LLC
Vero Beach, Florida 32964

PROJECT EDITORS
Teresa and Ed Sikora

PHOTO CREDITS
Gibbons Photography
PhotoDisc
Walt Burkett, Photographer

ILLUSTRATIONS
Kathleen Carreiro

EDITORIAL SERVICES
Pamela Schroeder

Library of Congress Cataloging-in-Publication Data

Dalton, Cindy Devine, 1964–
  Force & Motion / Cindy Devine Dalton.
    p. cm. — (How can I experiment with?)
  Includes bibliographical references and index
  ISBN 978-1-58952-012-7 (Hard Cover)
  ISBN 978-1-58952-017-2 (Soft Cover)
  1. Force and energy. 2. Experiments—Juvenile literature. 3. Motion—Experiments—Juvenile. I.Title

QC73..4.D35 2001
531.621   01-019052

Rourke Publishing
Printed in the United States of America, North Mankato, Minnesota
042010
042010LP-A

Force: something, such as a push or a pull, that changes the speed or direction in which something moves.

Motion: the act or process of moving.

Quote:

"Efforts and courage are not enough without purpose and direction."

-John F. Kennedy

Table of Contents

What Does Work Have to Do With It?

To learn about **force** and **motion**, you must first learn about work. Are we talking about homework? No way! We're talking about **mechanical** work.

Work can be described by the formula, force times distance. Sound confusing? It's not. Picture this: you have a box on one side of the room and you would like to move it to the other side. It's going to take some work. But how much? Just multiply the force needed to pull the box times the distance you pulled it. You've just figured out work!

If you'd like to try this at home or school, ask your teacher for a spring scale. Attach it to the object you want to move and pull it. It will tell you how much force times how far you moved it. You have just experimented with work!

Work = Force x Distance

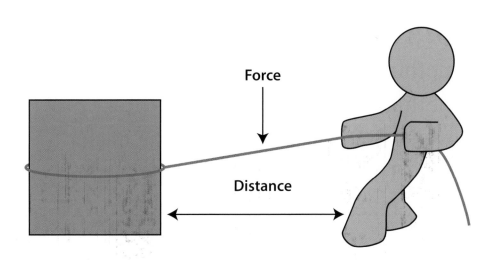

What Is Force?

When you think of force, do you think of how hard you have to work? If you wanted to tackle someone at a football game, how hard would you have to hit him to knock him down?

Why does it take a truck longer to stop than a car? Have you ever thought about how long it would take a ship or a train to stop? The answer to these questions has something to do with **mass**. The more mass something has, the more force it takes to move it. The less mass it has, the easier it moves. The faster an object moves, the more force it takes to stop it. If something has a lot of mass and is moving fast, like a train, it takes a lot of force to stop it.

Think of how much work it takes to tackle someone. That is the amount of force needed to change an object's movement.

Machines Help Us Do Work

There are a lot of machines in the world and they all do one thing. They help us do work easier. Has your mom or dad ever asked you to take something to your neighbor? Were there too many things to carry on your bike? You might pull your wagon behind your bike to carry your things, right? You were using a complex machine to help you do the work easier.

The lever, pulley, wheel and axle, inclined plane. and the wedge are examples of simple machines, you use them every day.

You use simple machines every day. Without machines, the work you do would be much harder.

The Lever

Simple machines have been around for thousands of years. A simple machine may or may not have moving parts. One simple machine without a moving part is a lever. Have you ever dug a hole with a shovel? That is one type of lever. Levers can also have moving parts, like a seesaw. A seesaw has a **fulcrum** in the middle. A fulcrum holds and **balances** the weight. One person sits on each end and each end moves up and down. Can you think of other levers? A teeter-totter, a wheelbarrow, the claw end of a hammer, and even piano keys are all levers.

Levers are used in many ways. You cannot see them, but piano keys are levers.

The Pulley

The pulley is a simple machine used to lift objects. Do you have blinds on your windows? When you pull on one string, the blinds come up. You are using a pulley. A flagpole also uses a pulley system to raise a flag. Sometimes things that need to be lifted can get very heavy, such as a bucket of water in a well. A pulley reduces the work you need to lift things.

Pulleys are quite amazing! Let's say you wanted to lift something that weighed 100 pounds (454 kg). Attach one pulley to the object and you would still have to pull with the force of 100 pounds.

Pulleys really do make our work easier.

If you'd like to try this at home or school, ask your teacher for a spring scale. Attach it to the object you want to move and pull it. It will tell you how much force times how far you moved it. You have just experimented with work!

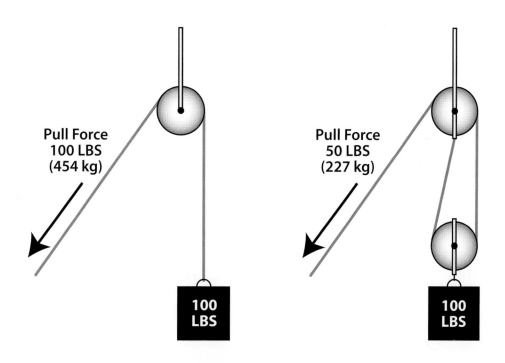

Pull Force
100 LBS
(454 kg)

100
LBS

Pull Force
50 LBS
(227 kg)

100
LBS

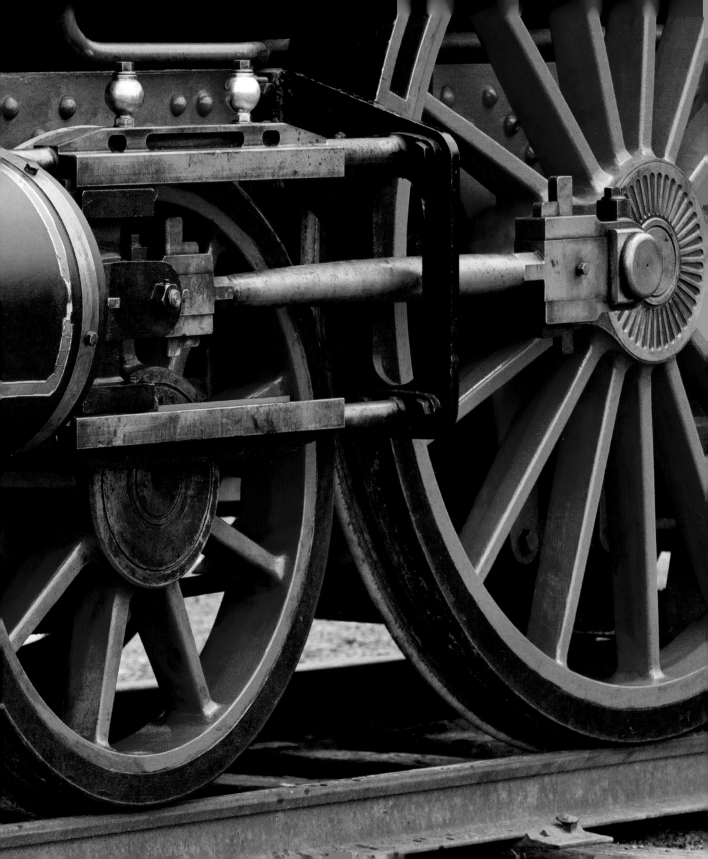

The Wheel and Axle

Do you think of a bicycle, a doorknob, or a clock as machines? These are great examples of wheels and axles. An axle is a **shaft** that is attached to the center of a wheel. When you turn the axle, the wheel moves forward. Using a wheel and axle, you can move heavy objects over a distance more easily. Imagine how hard it would be to move a car without wheels!

People in the past learned that they could move things more easily using wheels and axles. There is proof that wheels and axles were used as early as 3500 B.C. People used them on carts, potter wheels, and windmills.

Wheels and axles aren't just for cars. Doorknobs, clocks, and umbrellas all use wheels and axles.

Inclines Planes

Have you ever tried to walk up a steep hill? If you have, you know how difficult it can be. Picture a mountain with one side that slopes up gradually and another side that goes straight up. Which side do you think would be easier to walk up? If you guessed the sloped side, you're right! That's how inclined planes make our lives easier. You may have seen an inclined plane or ramp on the back of a moving truck. Movers use the ramp to load things onto the truck to make their jobs less difficult.

Inclined planes help you climb up easier because the work is spread out over a longer distance. You move up a little bit at a time, instead of straight up. Playground slides and skateboard ramps are good examples.

Imagine not having an inclined plane for your sled! It would be a hard climb and a dangerous fall.

The Wedge

Another type of inclined plane is a wedge. A wedge is two inclined planes put together. They can form a **blunt** edge or a very sharp edge. You may use a blunt wedge as a doorstop. Examples of sharp wedges are axes and knives. Wedges were used by the Egyptians as far back as 287 B.C.

A knife is a simple wedge that is used around the house every day.

Complex Machines

Complex machines are made of two or more simple machines. A car is a good example. It is made of wheels and axles, levers, and pulleys. All machines have one thing in common. They make work easier by helping us use less force. This is called **efficiency**.

Complex machines are two or more simple machines put together. Cars and bikes have wheels and axles, levers, and pulleys to make them work.

23

Force Made Simple

Simple and complex machines were invented to help make work easier. Easier work really means using less force on an object to make it change. If you think of all the machines around you, they all do one thing. They help us accomplish a task. Computers help us gather information and communicate easier. Automobiles help us travel farther and faster, and tractors help us grow food. What other machines can you think of that make your work easier?

Simple parts of an umbrella make up a complex machine.

Hands on:
A Force and Motion Experiment

What you need:

- A stool or chair that can spin
- A partner
- Adult help

Try This:

1. Sit in a chair with your arms outstretched.

2. Have your partner spin you. Then have that person let go and move away.

3. Quickly pull your hands against your body. Notice that you turn faster. Be careful! A very fast spin can make your chair tip over. Also, you may be dizzy when you get up.

Start with your arms stretched out.

Pull your arms in and see how much faster you spin.

What Happened?

Isaac Newton told us that objects stay in motion until a force changes the motion. A turning object will keep turning at the same speed unless another force acts on it. The farther the mass of the object is from the center of the spin, the slower the spin will be. Think about the experiment. You spin slowly when you hold your arms away from you. When you bring them in, you spin faster.

Now you know how ice skaters spin so quickly. Divers and gymnasts use this law of science, too. They speed up their spin when they change from a laid-out position to a tuck position!

Sir Isaac Newton helped the world understand basic principles of science.

Glossary

balances (BAL-en-sez) — makes even

blunt (BLUNT) — having a thick edge or point

efficiency (ih-FISH-en-see) — working very well

force (FORS) — a push or pull that changes the speed or direction in which something moves

fulcrum (FOOL-krem) — the support that a lever rests on

mass (MAS) — the amount of matter in something

mechanical (meh-KAN-ih-kel) — force used on an object

motion (MOH-shen) — the act or process of moving

shaft (SHAFT) — a long bar or rod

Websites to visit

www.Kapili.com
www.exploratorium.edu
www.littleshop.physics.colostate.edu

Index